CYBERSECURITY LESSONS FROM THE COVID-19 PANDEMIC

How to keep yourself and your family safe
when working, learning, and socializing online

Lysandra E Capella

CONTENTS

Title Page   1

Foreword   5

Chapter 1   Basic Hygiene Matters

Chapter 2.   Contact Tracing

Chapter 3. Flatten The Curve

Chapter 4. It's In Your Hands   23

Afterword   25

FOREWORD

T he real world is fragile. It is frightening to see how a virus can cause such massive disruptions across the globe. The digital world is fragile as well, as past cyberattacks have shown us. Many of the concepts and much of the language of cyber security are borrowed from the medical world's continuous battle against microbiological threats. The type and depth of response needed to battle a pandemic is similar to the level and depth of response needed to ward off cyberattacks.

Our digital networks can be attacked by malicious viruses that can propagate and spread, causing widespread network infections. To defend, we try to identify the contaminated software or device and then we isolate it from the rest of our network. We teach our citizens to protect themselves with anti-virus software and to be very careful when electronically touching (clicking on) things, because they could get infected.

Every crisis teaches us life lessons and this one has also taught a few lessons that can be related and taken note off in both the worlds – medical and digital. To counteract COVID-19, there are simple lessons that people around the world are learning about hygiene, applying these lessons to viruses and cyber viruses they encounter every day.

COVID-19 has changed how people work, and cybersecurity needs to reflect this new normal. The challenge here is not new technology, but old mindsets.

Think of this as a refresher on hygiene and a lesson on cyber hygiene!

CHAPTER 1   BASIC HYGIENE MATTERS

Attackers thrive in the places we cannot see, in much the same way that microbes hang on wherever we do not spray the disinfectant. The current strain of Coronavirus may be new, but it still exploits the same attack vectors that humans have had since prehistoric times – make one victim cough and depend on poor hygiene to infect the next person. Modern humans have the ability to stop these diseases, because we have water and soap, but they are only effective if we actually use them.

The mundane nature of the best counter to Covid-19 – just wash your hands – is a reminder that basics are still our most important line of defense.

Microbes have to obey the laws of biology – they cannot just teleport from person to person, they need a way to get between them, and at least for airborne pathogens, it creates a chain that we can break with something far less costly than a super drug.
At the bottom of so many of breaches is a failure to patch and otherwise update systems. That's the cyber hygiene equivalent of washing your hands.

Lesson I.   Wash Your Cyber Hands!

L ike the simple act of hand-washing, a culture of cyber-awareness does not have to be complicated or expensive. Just like you'd put a bandage on a cut to keep the wound from becoming infected, ensuring that any software you use has the latest patches can help keep bad actors from infiltrating your system.

KEY TAKEAWAY

1) If you receive notice from a piece of software you use that a patch is available, be sure to install it as soon as possible. Patches often resolve weaknesses and security vulnerabilities within products. Patching lessens the risk that a hacker can take advantage of a previously existing vulnerability.
2) Activate automatic software and application updates on all your devices. This rule applies to almost any technology connected to a network, including not only your work devices butInternet-connected TV's, baby monitors, security cameras, home routers, gaming consoles and more.

Viruses and other infectious diseases spread when humans connect with each other. As COVID-19 has proven, our interconnected lives have been a powerful vehicle for transmission.

The same can be observed of threat actors. After identifying a single point of access, like an unpatched system, they can move laterally across interconnected networks to find their intended target and fulfill their nefarious intentions. 2017's NotPetya attack, which was the most costly cyber-attack in history, is a classic example of malware taking advantage of unpatched systems to wreak havoc.

Lesson II. Disinfection of common and high-traffic areas

C leaning is top of mind during the coronavirus pandemic. The coronavirus is known to survive undetected on surfaces such as door knobs, grocery store carts, elevator buttons, and packaging for hours or even days. With so many transmission points, care is needed to mitigate risk. You might regularly wash everything from your hands to your doorknobs. It's a good idea to make sure your cell phone is on your cleaning list.

This good hygiene also extends to the software inside your phone. Clean software helps protect your data. Antivirus software and regular program updates all help to keep viruses, malware, and cybercriminals away from your system. Mobile devices are often a target for malware, which can infect mobile devices. That might occur via a compromised app, phishing email, or SMS text message. Common types of mobile malware include ransomware, trojans, and spyware.

KEY TAKEAWAY

There are several ways to keep your mobile operating system and data

secure :

1.   Delete any unused apps. Unwanted apps can take up space, drain battery, and slow down the device. Plus, old apps can let in malicious software, especially if the apps haven't been updated with the latest security patches. If you want to keep an app, be sure to update it regularly.

2.   Review the apps already installed on your phone, you might also check for excessive permission requests or settings.

Lesson III. Reduce Entry – Password Hygiene

C oronavirus is a good example of crisis-driven attention to a neglected area. Normally, we fly around visiting busy places, shaking hands, and generally behaving as if the outside world was not out to get us.

During the coronavirus crisis, the proliferation of hand sanitizers everywhere in workplaces, supermarkets and other places has made it obvious to splash your hands regularly without actively thinking about it. Just being there will be part of the habit. And if they are strategically placed, e.g. in the middle of an entrance, there is a strong social pressure to use them. You do not want to seem irresponsible to others.

Measures such as these, make the desired behavior the most obvious thing to do in the situation. And that is exactly what is needed to be achieved with digital security behavior. The longer we live and behave in a changed environment, the greater the likelihood that the good habits will stick. They will become a natural part of what we do every day, thereby redefining who we are. Only good habits separate the weakest link from a security hero.

KEY TAKEAWAY

Covid-19 has many people changing the devices they work on, and therefore losing all their saved passwords. To help solve this problem, use a password manager like Dashlane or LastPass to keep track of all credentials. They log in automatically and use complex encryption to

keep passwords out of cybercriminals' hands. Some relevant password managers include Dashlane, Bitwarden, LastPass, Keepass and many more.

Lesson IV. Be Careful when electronically touching (clicking on) things

J ust like not washing your hands after touching things at a public place could cause you to become infected with COVID-19, it only takes a momentary lapse in judgment for you to click on a link and potentially infect your computer.

People not practicing basic hygiene endanger us all eventually. Likewise, people not practicing basic cyber hygiene endanger their organizations. People may be even more important than the best technology defense. As we are operating more cautiously while understanding the risks associated with COVID-19, we should take the same precautions when it comes to cybersecurity.

Like during COVID-19, it's important for organizations to provide on-going security awareness education to make sure the needed protection for an attack is provided. The added stress many people felt during the pandemic has made them more prone to social-engineering attacks.

Cyber criminals are well aware of this, which is why Remote Working Cyber Attacks have risen during the shift to working from home. With any situation where infection is a possibility, a healthy amount of skepticism is always warranted.

KEY TAKEAWAY

1) Be wary of emails coming from unknown sources, particularly if the requester is asking you to click on a link or an attachment. When in doubt, do some Googling, let your IT/security team know or pick up the phone and call someone to ask if their request is valid.
A fake site might look just like the real thing, but the URL won't.

2) Use Ad Blockers to skim out fraudulent shady information, available through your browser's store.

CHAPTER 2.
CONTACT TRACING

V arious approaches to detect infected people are key in handling the crisis. In the fight against disease, it's increasingly clear that the difference between countries that have better or worse outcomes comes down to who can test the most.

They can see where the disease really is and get ahead of it. Digital security is the same. The decisions made are only as good as the data available.

Testing has proven to be an essential tool in fighting the COVID-19 epidemic. Infection proliferation through pre-symptomatic and asymptomatic people has caused geometric growth of cases and overwhelmed health care systems around the world. Early testing could have slowed the case growth rate and reduce the severity of individual cases. Lower case levels and death rate percentages in countries that used early detection demonstrate its effectiveness in managing major outbreaks.

Lesson V. Block Entry –Test, Test, Test

M any countries effectively traced infected areas and people. In the extreme cases, apps that tracked and located infected persons were made. Similarly, using cyber forensic methods and techniques, infected machines can be traced and identified.

Another important similarity is the need for an antivirus vaccine. Classic antivirus computer solutions work in a way that is similar to how our immune systems fend off viruses. The immune system in the body saves a small section of the virus and uses that as a way to identify infected cells, which it then destroys. Antivirus systems actually do the same thing by saving a small piece of the virus and create files to identify virus-infected files.

KEY TAKEAWAY

1) Install reliable antivirus software, such Norton AntiVirus, Kaspersky, Bitdefender, and Windows Defender, and keep it updated. These programs offer real-time monitoring for viruses, malware/spyware, and ransomware.
2) Scheduling regular scans.
Think of it like getting a checkup, just more often. Be sure to keep your antivirus software updated!

Lesson VI. Early Detection

For real-world diseases, contact tracing is used. If one person is a carrier, immediately their contacts are tracked down, tested, and quarantined if necessary.

The digital version of the challenge is much harder because computers communicate across a network in many different and shifting directions, comparable to having every person on earth flying country to country every day.

Like COVID-19 testing, the effectiveness of early detection is directly related to how quickly problems are uncovered. Visibility during the normal course of business will enable it to manage its assets. Once there is a crisis (or attack), visibility brings clarity into what is happening, where it is occurring, and the extent to which the business will be affected.

In addition, the ability to answer specific questions can help coordinate the response, just like governments and health officials need to be able to monitor disease hotspots and high-risk areas.

For example, for many organizations during NotPetya (targeted ransomware) attacks, it would have been helpful to have a map of all their SMB connections – before they were compromised. Rapid identification of compromised assets offers analogous benefits in managing major cyber-attacks.

Typical households have numerous smart devices such as cellphones, tablets, computers, Alexa, Smart TVs, etc. All of these devices can affect a network. Home Internet access is typically less secure than a business network. As more employees work from home, home networks must be secured as much as possible.

KEY TAKEAWAY

1) Change the default password on your router as well as on any other home network devices.

2) Do not connect to any unsecured or unknown Wi-Fi networks; only connect to Wi-Fi networks secured with a password.

3) In configuring your home network, secure it with a unique password, and ensure that it is protected using WPA2, Wi-Fi Protected Access 2.

4) Create one network for personal use, and the other for professional work.

CHAPTER 3. FLATTEN THE CURVE

E very country has limited resources to handle a pandemic like COVID-19. No health care system is designed to handle enormous numbers of infected people at the same time.

It was clear from the start that COVID-19 was deadlier to certain demographics, such as those with immunodeficiencies and elderly citizens. With this in mind, many countries have warned at-risk populations to avoid travel and stay at home as much as possible. Reducing contact with potentially infected people is central to protecting important and at-risk populations.

Too many people becoming severely ill with COVID-19 at roughly the same time could result in a shortage of hospital beds, equipment or doctors. On the other hand, if that same large number of patients arrived at the hospital at a slower rate, for example, over the course of several weeks, the line of the graph would look like a longer, flatter curve. In this situation, fewer patients would arrive at the hospital each day. There would be a better chance of the hospital being able to keep up with adequate supplies, beds and health care providers to care for them.

When the threat really does land home, we need to know how to respond effectively. In a pandemic, this can mean hospitalizing the most vulnerable population, the very young or very old, and asking the healthier to self-quarantine unless their symptoms become dire.

Cybersecurity resources are no exception. Widespread attacks can create spike in the resources needed for response. Risk based network segmentation, proactive isolation, and prioritization on critical assets can help flatten the curve and lower the impact which leads to faster recovery times.

With a good updated inventory, the most important and vulnerable applications can be prioritized. Other lower priority compromised systems may not be worth saving and just rebuilt from bare metal using automation. Consider the spread of malware or an attacker moving within an organization; it's better to lose a handful of systems while you put monitoring and remediation in place to harden the rest.

Lesson VII. Social Distance

Social Distance is used to stress the importance of maintaining physical space when in public areas. Wearing a face mask or covering when at home and whenever not around people who are not members of the same household. Maintaining at least 6 feet of distance, avoiding crowded places, particularly indoors, and events that are likely to draw crowds.

In practicing digital distancing, the basic realization is that if you or someone close to you falls victim to a cyberattack on your home network because you failed to take the proper precautions, you can become a vector for a larger intrusion affecting countless others beyond yourself.

The implication is that just as is the case with social distancing, we bear a social responsibility to those around us to do what we can to remain secure.

1) Use a Virtual Private Network (VPN). A VPN obfuscates all of web traffic, both by encrypting the data transferring back and forth from your system as well as masking your location and IP address. The result is that it becomes more difficult to snoop on what you're doing online, giving you a solid layer of security against would-be attackers. There are a number of VPNs to choose from, with some being free and others requiring a subscription while offering enhanced protection.

It is common practice not to share your personal details online. In the same way you wouldn't walk up to someone coughing at the supermarket, you need to be careful with who you interact with online. This element of privacy is akin to using a mask, something that everyone is doing now. No matter if you are on-line or on the street, keeping your face hidden will keep you safe from all sorts of troubles and worries. Masking your private data is always the smart thing to do.

Lesson VIII. Shelter in Place

By sheltering in place, we impede the coronavirus' ability to move laterally around the population. By avoiding close contact with anyone showing symptoms of respiratory illness such as coughing and sneezing, maintaining a distance of at least 2 meters (6 feet).

One of the greatest defenses against ransomware attacks is to have solid and recent back-ups in a segregated environment that would not fall victim to the same hack on your corporate system.

Sometimes, despite best efforts at hygiene, you can get infected. The same goes with your digital devices, Loss, damage, technical malfunction, sabotage, and theft can never be fully prevented, so make sure you have a reliable system for backing up your data.

KEY TAKEAWAY

The quickest, easiest way to back up your business PC from home is to use a cloud backup service. Ones to consider are Acronis, Carbonite, and IDrive or Google's Backup and Sync.

CHAPTER 4. IT'S IN YOUR HANDS

C OVID-19 is not the only risk with the ability to quickly and exponentially disrupt the way we live. The crisis shows that the world is far more prone to disturbance by pandemics, cyberattacks or environmental tipping points than history indicates.

Our "new normal" isn't COVID-19 itself – it's COVID-like incidents. And a cyber-pandemic is probably as inevitable as a future disease pandemic.

The time to start thinking about the response is – as always – yesterday.

Let's summarize what we learned from Covid-19

1. Wash Your Hands Regularly
Keep your systems and applications patched and updated.
2. Disinfection of common and high-traffic areas
Keep your mobile operating system and data secure
3. Reduce Entry
Use a password Manager to protect your passwords
4. Don't Touch Your Face
Be wary of emails coming from unknown sources
5. Block Entry
Install reliable antivirus software
6. Early Detection

Secure your home network
7. Social Distance
Use a VPN whenever possible
8. Shelter in Place
Perform regular Offline Backups

No matter if we look at pandemics or cybercrime, humans are the only link when responding to changes in the risk landscape.

Handling new risks means that humans must behave differently from status quo. The pandemic can seem overwhelming, but in truth, every person can help slow down the spread of COVID-19.

By doing your part, you can make a big difference to your health, and that of others around you.

AFTERWORD

Stay safe & keep washing your hands!